Little Fingers BIG Science

This book belongs to ______________________________

Published in Virginia, USA.

ISBN: 978-1-957753-03-4

Turtle Trails Publishing

www.TurtleTrailsPublishing.com

Do You Have a Heart?

Carla Mae Jansen

Do you
have a
heart?

Yes, I have a heart!
(cat)

Do you have a heart?

No, I do not have a heart!
(jellyfish)

Do you
have a
heart?

(frog)

Do you
have a
heart?

(sea stars)

Do you
have a
heart?

(bald eagle)

Do you
have a
heart?

No, I do not have a heart!

(sea cucumber)

Do you have a heart?

Yes, I have 3 hearts!
(octopus)

Do you
have a
heart?

(sea anemone)

Do you
have a
heart?

(clown fish)

Do you
have a
heart?

(fox coral)

Do you
have a
heart?

(earthworm)

More About

Hearts pump blood to our bodies. All mammals (like cats, dogs, monkeys, opossums, and rabbits) have a heart with four chambers.

Birds also have hearts with four chambers.

Fish have hearts with only two chambers.

Hearts!

Amphibians like frogs and salamanders have a heart with three chambers. Most reptiles (like snakes) also have hearts with three chambers. (Crocodilians are the only reptiles with four-chambered hearts.)

An octopus has three hearts: one pumps blood to their body and two pump blood to their gills!

Earthworms have five very simple hearts called "aortic arches" to pump blood down their bodies!

More About

Jellyfish, sea anemones, sea stars, sea cucumbers, and corals do not have blood, so they do not need hearts to pump blood. Animals with hearts need blood to carry oxygen to their bodies. Jellyfish, sea anemones, sea stars, sea cucumbers, and corals pull oxygen directly out of the water. They need water to pass over their bodies, but they do not need hearts or blood.

Hearts!

Mammals, birds, reptiles, amphibians, and fish have "closed" circulation systems, which means that all the blood stays inside blood vessels and the heart. Insects have an "open" circulation system, which means the "blood" and other fluids just slosh around inside their bodies. Insect "hearts" are very simple tubes that make sure the fluids keep moving. Those fluids are important, but they do not deliver oxygen. Insects have a different system of tunnels for oxygen and other gasses.

Heart

Human children hearts usually beat around 70-115 times a minute!

Whales have the largest hearts!

Cat hearts beat 140-220 times a minute!

Your heart is about the size of your fist.

Fun Facts

The first open-heart surgery in the United States was done in 1893 without x-rays by Dr. Daniel Hale Williams!

Your heart rate is controlled by electricity!

Pygmy shrew hearts beat 1200 times a minute!

Which animals do NOT have a heart?

Meet the Author:

Carla Mae Jansen loves animals and learning all about them! She is a teacher (MA Science Education), mom, rock-collector, and author.
You can reach her at
Carla@TurtleTrailsPublishing.com.

Also by the Author:

Did you love this book? Please leave a review!

Why do skeletons
avoid medical care?

Because their hearts
aren't in it!

Made in United States
Troutdale, OR
11/03/2026

56844688R00021